D1146970

PAVE IT

Dek Messecar
Series Consultant Editor: Bob Tattersall

CONTENTS

COLLINS

Introduction

A firm, well-drained outdoor surface is a practical requirement on any ground where there is heavy traffic, whether of feet, wheels, or hooves. Paving of some kind is used to provide such a surface. Mud or dust, puddles and pot-holes, trampled soil or bare patches on grass, especially at exits and entrances, are all signs that paving may be needed.

Good paving lasts a long time: appearance as well as practicality may call for consideration. The colour, texture and pattern of paving, whether blending or contrasting, should be in keeping with the character of the house and its environment. Paving can enhance the appearance of a garden, defining areas and creating interest. On a terrace or patio it becomes an extension of the household's living area, reflecting the life style of its occupants.

The techniques of paving are not difficult, but work – a certain amount of time, physical strength and sometimes stamina – is needed. Small jobs may perhaps be managed by one person, but the support of at least one helper makes most jobs easier and more enjoyable, and can occasionally be essential.

Before making any preparations it would be a good idea to read through this book to give yourself an overview of methods, materials and equipment. Catalogues, price lists and visits to suppliers will provide invaluable guidance on the cost and appearance of materials.

Most important of all, successful paving depends on good advance planning – the careful measurement of distances and quantities, attention to layout and preparation, and a schedule that allows time for processes such as hardening as well as for the work itself. Patience and forethought are the best workmates for a professional-looking finish.

Left *During the summer months your patio can become an attractive extension to your home.*

Above right *This old stone path has a charm which is hard to achieve with new materials. One solution is to buy old bricks–they may take some time to find, but your effort will be rewarded.*

Right *A purple lavender hedge blends well with the grey gravel drive it edges, and helps to prevent the gravel from spreading into the garden.*

Top *Granite sets make an attractive and hardwearing driveway – they have been bedded down in mortar for extra durability. Note the stone slab border which adds style.*

Above *York stone flagstones of irregular size have been laid with gaps between, which have been planted with interesting low plants. This would be suitable only for a path or area of paving which did not get used very much.*

MATERIALS

There is a wide range of materials that can be used for paths, patios or drives. Among the most common surfaces are paving slabs, concrete, bricks and gravel. The choice of which one to use depends on how the material blends with the surroundings, how difficult it is to lay, and, of course, price. Here is a list of possible materials.

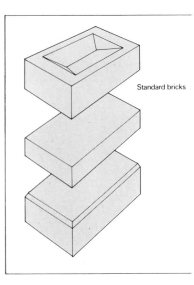

Standard bricks

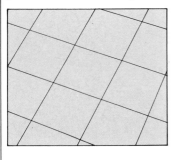

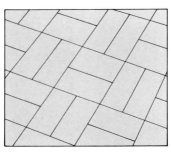

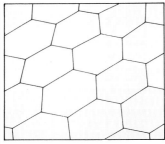

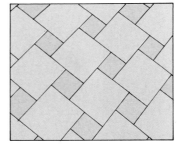

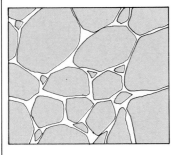

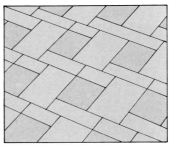

Paving slabs are made of pre-cast concrete and come in varying shapes, sizes, colours, patterns and textures. They are mainly used for paths and patios, although the strong, hydraulically pressed type can be used for drives.

Plain square or rectangular slabs are the least expensive and simplest to use, but can look dreary if used to cover a large area. However, you can combine different sizes, colours and shapes to create interesting effects. Broken slabs (*crazy paving*) look less monotonous.

Standard bricks or thinner brick pavers can be used to make paths, drives and patios. It is important to use frost-proof bricks or the thin bricks made especially for paving.

Although bricks are more expensive and time-consuming to use than paving slabs they can be an attractive alternative if a traditional, patterned look is desired. An added advantage is that brick paving goes well with a brick-built house. A popular pattern used with bricks is the herringbone pattern (see page 11).

Paving blocks are smaller, thinner versions of paving slabs. They are made of concrete in many colours and shapes.

Stone in several forms is used for paving, either alone or with other materials.

Cobbles are small, rounded stones which can be laid on

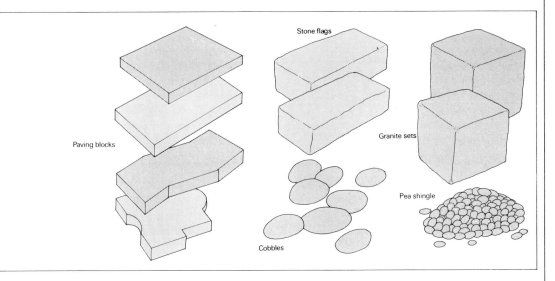

Stone flags

Paving blocks

Granite sets

Cobbles

Pea shingle

edge or flat on a bed of mortar. They are often used to fill in areas between other types of paving.

Stone flags are slabs of sawn or rough-hewn stone that can be used to pave an area or laid in a single row to make a path.

Granite sets can be brick-shaped or cubic and are the traditional surface for city streets. Although expensive they give an extremely durable finish.

Concrete is a mixture of sand, cement and aggregate which sets to form a very hard surface. The thickness of concrete required will depend on the intended use; a drive will obviously need a greater thickness than a path. Concrete can look dull but is cheaper than any of the materials described above. It can be made brighter and more colourful by using powdered pigment in the

mixing process or by staining after laying.

Concrete will crack if laid over too large an area. Large expanses should be made up of small sections (see page 14). Concrete can be mixed by hand or with a hired cement mixer. Large quantities are best purchased ready-mixed and are delivered by lorries.

Asphalt, sometimes called *macadam,* is a mix of bitumen and fine gravel laid on a bitumen emulsion base. Tough and waterproof, it is widely used to resurface drives. It is relatively cheap but messy to use – wear overalls to protect clothing.

Gravel is sharp-edged chippings of crushed stone and varies in size and colour depending on its origin.

It is usually laid loose, although it can be fixed by laying it on a bitumen emulsion base. A retaining

edge, usually of kerbstone, brick, or treated timber, is necessary to keep loose gravel within the area it is meant to cover. Gravel is used mainly for paths and drives. It is cheap and easy to use but requires frequent raking to maintain an even spread. Gravel should be laid to a depth of about 25mm.

Because gravel has a tendency to move it is unsuitable on steep slopes.

Another loose paving material is *pea shingle* or *pebbles.* These are small beach stones and are sold in graded sizes.

Other materials are needed for underpinning and binding the paving. *Hardcore* is a mixture of broken bricks, concrete, and stones. *Ballast* and *coarse aggregate* are mixtures of sand with gravel or shingle for levelling the sub-base or making concrete. *Sand* and *cement,* and sometimes *pva adhesive,* are also needed.

TOOLS

Not all the tools listed here will be needed for every job. You may need to borrow or hire the heavier items required for big jobs. Other tools are simply pieces of timber cut to the right size and shape for a particular task. Some tools and working materials are not mentioned here in detail: a sharp knife, a saw, scissors and string, timber for pegs, battens, and shuttering, and polythene sheeting for protecting work from extremes of weather.

A *spade* is used to cut the turf, remove topsoil from the site and to dig out the sub-base to the necessary depth.

Shovels and *buckets,* at least two of each, are used for mixing concrete and mortar. One of each is kept dry and used for cement only. A *watering can* is also needed for mixing.

Marker pegs are wooden stakes driven into the ground around the edge of the sub-base site to show the level of the paving and the depth of the sub-base materials.

An *earth rammer* is used to consolidate the earth at the bottom of the sub-base and to pound the hardcore layer before laying the paving.

A *plate vibrator* is used for the same purpose and is also the best tool for bedding slabs, bricks and blocks on a dry sand bed.

A *garden roller* is used to level loose paving materials such as gravel and pebbles

and to lay asphalt when resurfacing paving that has deteriorated.

A *concrete mixer* can be hired inexpensively. It enables you to mix concrete more quickly and easily than by hand and in just the quantities you need.

A *tamping beam* is a tool you make yourself to suit the size of the concrete slab you are laying. It is used to consolidate soft concrete and to level the top surface with the formwork (see page 14). For a small path, cut a straight piece from 100mm x 50mm softwood. For larger bays add a handle to each end. Two people can then do the job standing up.

A *spirit level* – the longer the better – is used to measure gradients. It is

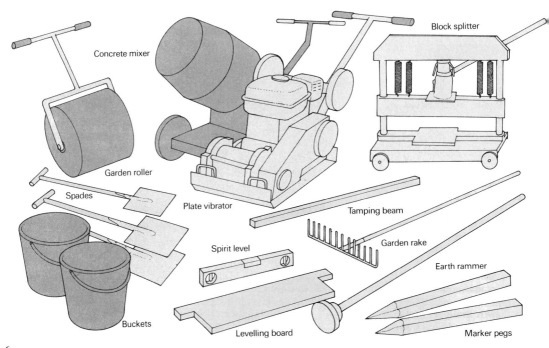

Concrete mixer

Block splitter

Garden roller

Spades

Plate vibrator

Tamping beam

Spirit level

Garden rake

Earth rammer

Buckets

Levelling board

Marker pegs

placed across the tops of the marker pegs to set the height of the proposed paving.

A timber batten can be used between pegs too far apart to be spanned by the spirit level.

A *bricklayer's trowel* is the best tool for working with mortar when setting edging blocks or kerbs.

Use a *wooden float* to spread a sand or mortar bed or to dress the surface of a wet concrete slab.

A metal *plasterer's trowel* may be used to give a concrete slab a very smooth surface.

An *arrissing trowel* has an angled edge and is used to trim the top edges of a concrete slab.

A *garden rake* is useful for spreading sand, mortar, concrete, ballast, gravel or asphalt.

A *levelling board* is another tool you make yourself. It is used to lay a bed of sand or mortar to a level that allows for the thickness of the paving.

A *bolster* and *club hammer* are used for cutting bricks, blocks or slabs.

A *block splitter* for cutting paving blocks, slabs or bricks can often be hired from suppliers of paving materials.

A *plumb line* and *household steps* are useful when planning and laying steps or paving at different levels.

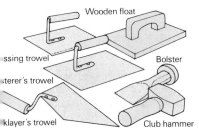

Wooden float

ssing trowel

terer's trowel

klayer's trowel

Bolster

Club hammer

Mixing concrete

Concrete for paving is mixed 1 part cement to 1.5 parts sharp sand and 2.5 parts coarse aggregate. This will make a strong, weather-resistant surface.

You can mix it with a shovel or hired concrete mixer, or it can be bought ready-mixed. The ingredients can be bought loose or bagged and ready to mix with water.

Mixed concrete is measured by volume in cubic metres, so find the amount you need by multiplying the area of the surface by the thickness of the slab.

To mix a small amount for setting kerbstones, the pre-bagged dry mix would be best, mixed on a board with a shovel in the same way as for mortar (see below).

If you are pouring up to about half a cubic metre a hired concrete mixer saves time and effort. Ask your supplier to calculate the quantities of the ingredients supplied loose and compare the prices.

For larger quantities, ready-mixed is preferable. There are two main considerations. The first is whether you can pour all the bays at one go – or will more than one delivery be necessary? How difficult will it be to get the concrete from the truck to the site and to lay it in the two hours it remains workable? A retarder can be added to the mix by the supplier to double the time before setting, if necessary.

Many suppliers sell concrete in all forms: seek a comparative quotation and expert advice.

Mixing sand and cement mortar

Mortar is used for binding and bedding bricks, blocks, slabs and stones. A bed of mortar can be mixed one part cement to four parts sharp sand. For spot bedding slabs or setting edgings, use the bricklaying mix of one part cement to three parts sharp sand. Always use a commercial plasticiser in mortar to make it easier to work with and to prevent cracking.

Mortar stays fresh for a maximum of two hours.

Use two buckets and shovels so that one of each can be kept dry and used only for cement. Measure the dry ingredients into a pile on the mixing board and mix thoroughly.

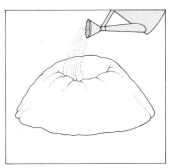

Make a crater in the top of the pile and pour in a little water from the watering can. Follow the manufacturer's instructions for adding the plasticiser.

Use the shovel to scoop the rim of the crater into the centre, turning over each shovelful. When the water is mixed, sprinkle a little more over the pile and keep turning it over until you have a stiff mixture that slides cleanly off the shovel.

Preparing the sub-base

All paving needs, for stability, a good foundation. The foundation, called a sub-base, is usually made of hardcore and ballast, and sometimes with concrete as well. For most non-specialist purposes a sub-base of a 100mm deep layer of hardcore with enough ballast to level the surface can be used for any situation and paving material. It is possible to reduce the depth on stable soil for paths and areas not bearing heavy loads. However, paving is meant to be permanent. Unless you have reliable advice to the contrary, it would be advisable to use the full sub-base described here.

Preparing the sub-base is the first part – and a large part – of the work that has to be done. But the first and most important part of the task as a whole is the thinking, designing and planning. You need a plan with measurements of the size and shape to be covered. Study the terrain and work out in advance how the paving is to be sloped for drainage and whether pipes, gullies or soakaways will be needed. This is especially important if the paving is to be close to a building. Look, too, for any features such as drainpipes, manhole covers or airbricks which may need special treatment (pages 12-14 provide guidance).

Use small pegs joined with string to mark the edges of your path or drive. Then you are ready to start on the sub-base.

and mark clearly the thickness of the paving material, the thickness of a sand or mortar bed (if appropriate) and the depth of the sub-base. Leave at least 250mm below this to hold the pegs firmly in the ground.

Drive them into the ground just outside the string. You will need one at each corner and others spaced not more than 1m apart along the edges. Site them wherever you will need to measure across them for setting the correct slopes and levels.

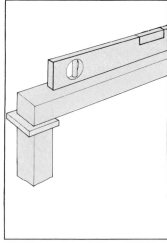

The tops of the pegs will be at the level of the top of the paving so, to set the required slopes, pack up one end of a timber straight-edge and use a spirit level to judge the height of the pegs. For instance, if the drive is 3m wide and is to have a crossfall (see page 12) of 1 in 40, the straight-edge should be packed up 75mm at the lower end. For a fall of 1 in 100 lengthways, use a packing piece 10mm thick for pegs 1m apart.

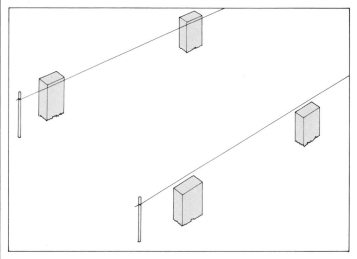

Make some marker pegs to suit the depth of sub-base required.

Use the top of the pegs as the level for the top of the paving

Dig out the area to the depths marked on the pegs, keeping the edges of the excavation vertical and in line with the string.

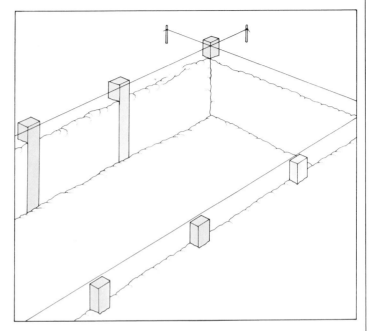

level mark on the pegs.

Keep ramming the hardcore down and levelling, adding more where necessary until the whole sub-base area is firmly compacted. A plate vibrator is the best tool for this: it saves time and usually does a better job than ramming by hand.

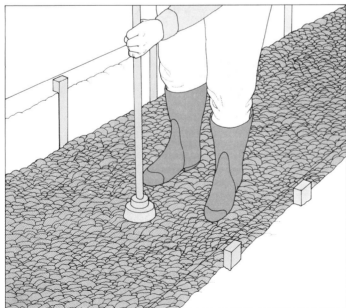

The last step is to 'blind' the hardcore with ballast, which is a mixture of sand and shingle. Spread the ballast over the hardcore to fill the gaps and level the surface ready for paving.

Consolidate the soil in the bottom of the excavation with an earth rammer and fill with hardcore up to the appropriate

PATHS AND DRIVES

Some form of paving is likely to be required on any path that is regularly used to and from or around a building. It is advisable, when laying a new path, to follow fairly closely the route that people are already using – usually the most obvious and direct one. If you do not, your carefully-laid new path may be ignored in favour of a more obvious route or 'unofficial' short cut.

A drive for vehicles, provided it is in good condition, is much less likely to be disregarded.

Paths in private gardens are also less vulnerable. They serve both practical and design purposes, and can be grand and formal or casual and winding, to suit and emphasize the character of the garden.

The concrete has been restricted to areas of hardest wear on this driveway, the area beneath the cars being surfaced with gravel.

Above left *Laying the bricks in a herringbone pattern has the effect of making the path seem interesting, rather than just long and narrow.*

Left *Paving stones set into a lawn are a practical solution when a proper path is not wanted or needed. Wherever your lawn gets undue wear, such as by a garden gate, you could consider inserting a paving stone.*

Above *Cobble stones have been bedded down well to provide a hard edge to this lawn as well as forming a path around the garden. Plants are allowed to grow over the edge of the border to soften the line between path and flower bed.*

Laying paths and drives

Paths and drives are laid in the same way and with the same choice of paving materials. However, as drives must support more weight, the sub-base must be given the full depth. The concrete, too, may need to be thicker and should include some reinforcement and expansion joints because of the larger area. The first step is to lay out the shape of the path or drive, taking account of any obstructions and slopes. It is important to give sufficient time and thought to the layout and the various slopes for drainage. Once paving has been laid, it is difficult to rectify mistakes later.

Planning

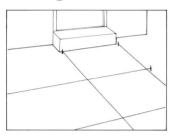

Use pegs and string to mark the edges. A drive needs to be 3m wide to allow for opening doors and for walking past parked cars. The width of paths may vary from a 300mm row of flag stones to 1.5 or 2m, depending on traffic and the effect you want to achieve.

Drainage

The next consideration is the way you will construct the paving to deal with rainwater.

All paved surfaces should have a slight slope or fall to ensure that rainwater runs off. A drive should slope away from the house, as should a path that runs along an exterior wall. A fall of 1 in 100 lengthways from the house towards the road is sufficient. There should also be a fall across the width (crossfall) of approximately 1 in 40. This

can be done either by sloping the entire surface towards one edge or by raising the centre slightly so the water runs off both edges. Choose the method which will lead the water in the best direction for your situation.

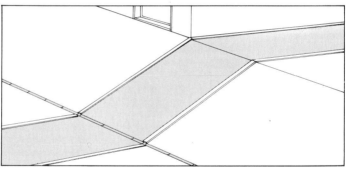

If your property slopes towards the house from the road you will need to install a drainage channel at the lowest point. The drive will then slope slightly upwards from the channel to the house.

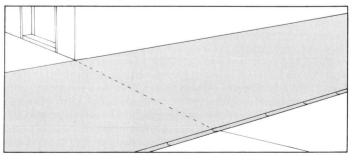

If the drive runs along the side of the house and the ground is sloping toward the house, you must install a drainage channel along the lower edge of the drive.

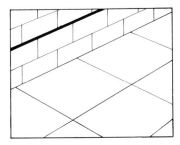

The top surface of any paved area adjacent to the wall of a building must be at least 150mm below the damp course.

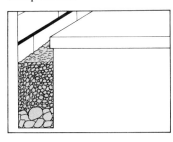

If this is not possible, a 75mm gap must be left between the wall and the paving. The gap should have hardcore at its base, and be filled with pebbles or gravel. The paving must slope away from the wall.

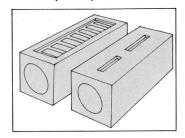

Drainage channels can be made from ready-made concrete channelling, either in the form of a gully or as enclosed pipe with slots or a grille in the top to allow the water through. These can be set into the concrete sub-base at a slope of 1 in 40. Alternatively, you can form your own gully in the wet concrete slab using a piece of pipe to shape it.

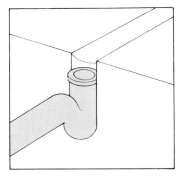

The water from the channels must be fed either to existing drains or led to a soakaway some distance from the house. A trap should be installed at the lower end of the channel to prevent clogging by leaves and silt.

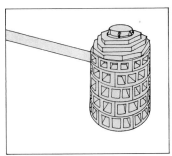

Soakaway drains are for surface water (not household waste) and can be purchased ready-made or you can make one yourself.

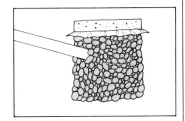

A home-made soakaway is a hole at least 1m square and 1m deep. The size depends on the surface area of paving and the type of soil. You should consult your local building regulations about the size and siting of soakaway drains.

The hole is filled with hardcore or rubble. The drain pipe is led into it near the top at a gentle slope. Then a sheet of heavy-gauge polythene is laid over the top and a 75mm layer of concrete is used to keep soil out. Topsoil and turf can be used to cover the concrete plug, hiding the drain completely.

Manholes

Manhole covers are another obstacle you may have to deal with. If your drive or path includes one, you may need to raise the level of the cover and its frame to the level of the new surface.

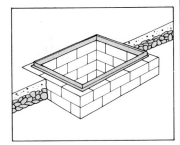

This can be done by chipping out the existing frame and

building up the level of the brickwork with extra courses of bricks and setting the frame in mortar as before.

If you wish to hide the manhole cover, there are now covers and frames available for the purpose. Some are for use with loose paving materials, such as gravel, while others are made to have concrete laid in the top of the lid. Follow the manufacturer's instructions.

Airbricks

If your drive or path is adjacent to a wall of the house or garage, check for any airbricks in the wall that are low enough to be covered by the paving. If there are any, set the height of the paving at least 50mm below the bottom of them. If this is not practicable there are two ways of dealing with them.

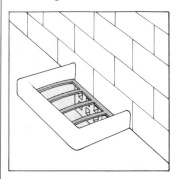

You can construct a low wall of brick or slabs or, in the case of poured concrete, simply leave an open well in front of the airbrick. A slight ridge around the top of the well is necessary to stop water from running in and it is advisable to add a metal grille to keep it clear of leaves. The bottom of the well should be hardcore

rammed down as for the rest of the sub-base. This will ensure that the small amount of rain that actually falls into the well can drain away.

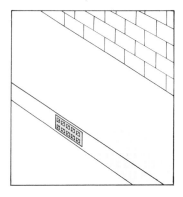

The other way to deal with an airbrick is to use plastic drainpipe set into the sub-base to duct air from the airbrick to a convenient vertical surface in the edge of the path or drive.

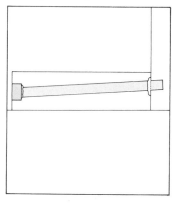

Arrange the pipe at a slight slope away from the wall in a trench in the sub-base. Chip out the airbrick with a cold chisel and club hammer and use sand and cement mortar to fix one end of the pipe into the hole and the other to the new airbrick.

Laying a concrete path or drive

A poured concrete path or drive needs timber *formwork* to hold the soft concrete and create neat edges. The top edges of the formwork are used as a guide when compacting and levelling the final surface.

As the formwork will be removed when the concrete sets, any boards at least 22mm thick or even strips of plywood can used.

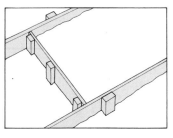

As well as as formwork around the edges of the path or drive, it will be necessary to divide the total area into *bays* with similar boards. There are several reasons for this. One is to enable the concrete to be poured, tamped and given a surface within the two hours maximum time that it remains workable.

A second reason is that slabs above a certain size must have expansion joints between them in order to prevent cracking. The general rule is that the thinner the concrete, the smaller the slabs should be. It is usual to make the length of the slabs about 1.5 times the width following the formula: 9 square metres maximum area for a 75mm thick slab, 12 for a 100mm slab, and 15 for a 150mm slab. However, even with helpers, a maximum size of ten square metres is advisable considering the time limit.

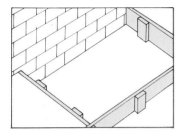

A third reason is, if a path or drive is adjacent to a wall, it will be necessary to divide it with formwork to make tamping possible. You can lay alternate bays and then, once they have set, remove the boards dividing the bays and lay the remaining bays by standing on the ones previously laid.

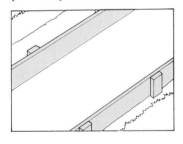

Fix the *shutter boards* (the wooden boards that retain the wet concrete) into the hole with the top edges flush with the marker pegs. If the surrounding ground level is to be higher than the slab, dig out a trench along the edge of the sub-base and support the shutterboards with stakes driven behind them. Use stakes to support the boards dividing the bays. Place the stakes on the outside of the bays that are to be poured first.

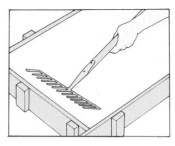

Pour freshly mixed concrete into the first bay. Use a rake or shovel to push it into corners and generally move it around and up to the formwork, ensuring there are no air pockets beneath the surface. Continue filling until the surface is 10mm or so above the edge of the boards.

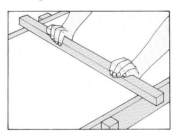

Use a tamping beam across the bay, repeatedly lifting and dropping it onto the formwork, moving forward 25mm each time. This compacts the concrete and levels the surface.

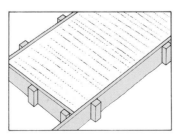

If an area sinks below the level of the tamping beam, toss a

shovelful of extra concrete on and tamp again until the entire surface is even.

You can leave the ridged texture made by the tamping beam as the final surface or smooth it further. The ridged finish is good for paths and drives with a pronounced slope, as it makes a good non-slip finish for both cars and pedestrians. There are several other finishes you can use.

Make a brushed finish with a stiff or soft broom held flat, brushing always in the same direction.

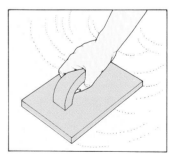

For a smooth finish, polish the tamped surface with a plasterer's trowel when the concrete has started to stiffen. Do not overwork the surface or you may make the aggregate sink in the mix and leave the surface too wet. It might then become weak and dusty later.

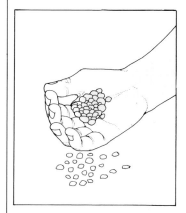

Stone chippings can be used to surface the fresh concrete. Sprinkle a layer on to the tamped surface and use the tamping beam again to press them into the concrete. When the concrete has hardened enough to hold the stones, spray the surface with a fine hose spray. Any loose stones can be brushed off with a stiff broom later.

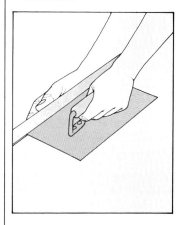

When the concrete stiffens, but before it hardens, round over the edges of the slab using an arrissing trowel along the join between concrete and formwork.

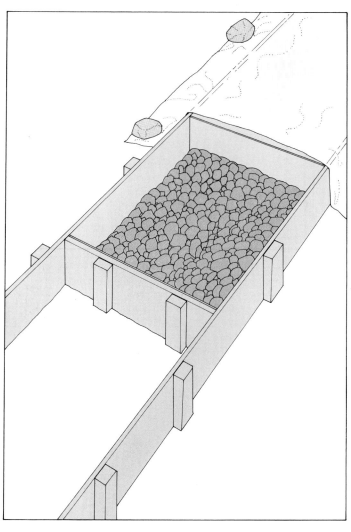

If you are laying a path that is accessible from either side, you can lay each adjacent bay by placing a strip of 3mm thick hardboard against the shutter board dividing the bays.

When the first bay is poured and surfaced, pour the second one part full and remove the shutter board, leaving the hardboard. Then continue to finish the second bay as the first.

Use a polythene sheet to cover fresh concrete for three days after laying to prevent its drying out too quickly. It can be walked on after this time, but full strength is not reached until after about 28 days.

Laying paving blocks and bricks

Paving blocks or bricks can be laid on a full bed of mortar or on a bed of sand. Both methods are strong enough to be used for a drive, but the sub-base must be very stable and well compacted – preferably using a heavy plate vibrator. For the sand method, a suitable vibrator should be used on the blocks after laying to compact the sand bed fully. For a small area or a path, hand compacting is sufficient.

Bear in mind that for the sand method the bricks or blocks will be laid tightly together. If you are using second-hand ones that may not be the same size, or if you want to include curves, the mortar method is best.

Blocks and bricks also require some form of edging to hold them together.

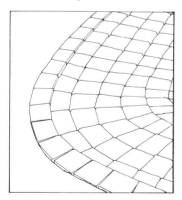

You can, if you wish, use the same blocks for edging, either on edge or on end. This can be useful if any edges are to be curved.

Concrete kerbs are less expensive and equally suitable.

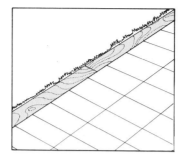

If the path is to be laid flush with a surrounding lawn, it is possible to use treated timber supported by stakes, although this is the least durable edging.

First, plan your layout as described earlier (see pages 12-14) for concrete paths and drives, taking into account the slopes, drainage, and any obstacles.

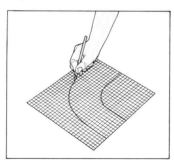

There are many patterns to choose from. Try to plan the paved area to accommodate the pattern you want using whole blocks to minimise cutting. Do this either by drawing a plan on graph paper, or by laying out blocks on the proposed site. Remember, if you will be laying them on sand, that they should be butted tightly together. On a bed of mortar

there should be joint gaps of about 10mm.

Set out the borders of the sub-base with pegs and string to the line of the paving blocks or bricks only. The edging will be laid after the sub-base is compacted. Make marker pegs with lines to show the thickness of the paving blocks or bricks, a 50mm layer of mortar or sand and 100mm of hardcore.

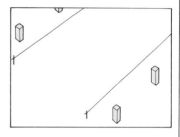

Drive marker pegs at the appropriate places but site them about 200mm outside the string lines. This is to allow for the concrete kerb that will support the edging.

Dig out the sub-base, fill it with 100mm of hard core and compact it well, preferably using a plate vibrator. Bind the sub-base with a layer of ballast and ram it down.

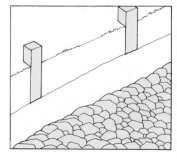

Lay a strip of concrete about 250mm wide along the edges.

The depth will depend on whether you want the edging to be flush or stand above the level of the paving. The concrete should be 75mm thick below the edging blocks or kerbs, so it may be necessary to remove some of the hardcore. Allow the concrete to cure for 24 hours before laying the edging.

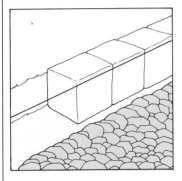

Using a generous bed of a stiff mix of sand and cement mortar, set the blocks, bricks or kerbs on the concrete along the outside lines of the string guides.

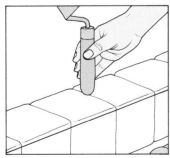

Tap the edging blocks down into the mortar until the top edges are level with the marker pegs, and allow a fillet of mortar to build up around the outer sides, adding some if necessary.

Lay the edging blocks tightly together unless they follow a curve. In that case allow a minimum joint gap of 10mm and wait 24 hours before pointing the joints with mortar.

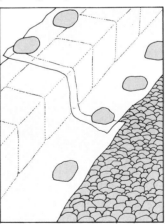

Cover the edging with polythene sheeting and allow it to cure for three days before laying the blocks.

Laying blocks and bricks on sand

The sand used for the bed should be as dry as possible.

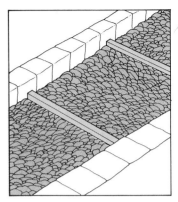

The easiest way to level the sand bed is to use two timber battens 60mm thick, cut to fit loosely between the edgings. The reason for the 60mm thickness is that the sand bed will be compacted to 50mm with the vibrator. If you are laying a path and are not using a vibrator, the difference will be less.

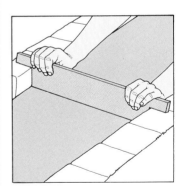

If you have edging on both sides, you can make a levelling board to rest on top of the edging to leave the sand about 10mm higher than it will be after compacting. So if your blocks or bricks are 65mm thick the sand should be filled to 55mm from the top of the edging.

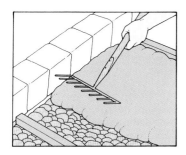

Shovel a thin layer of sand to cover the area and (unless you are going to use a levelling board) place the two battens about 1m apart. Use sand to pack them to the right level. Add more sand and spread it around with the back edge of a rake to ensure it fills all gaps and goes well into the corners.

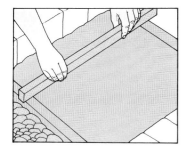

Use a straight batten to remove the excess sand, bringing the bed level with the battens.

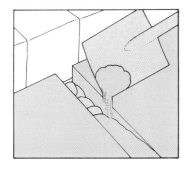

Remove the end batten and carefully fill the gap in the

same way as the first.

If you have easy access to the area you can lay the whole bed in this manner. Otherwise lay an area no larger than you need to begin laying blocks without walking on the bed.

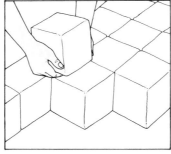

Start laying blocks in your chosen pattern in one corner of the bed. Place them touching each other and try not to slide them around to avoid disturbing the sand below.

It is best not to put any weight on the blocks, but if you have to kneel on them to reach, place a board to spread the load over as many as possible.

If you have easy access to use the vibrator, you can lay the whole site before compacting. However, it is a good idea to lay about four square metres and use the vibrator on them to check that the final level will be correct.

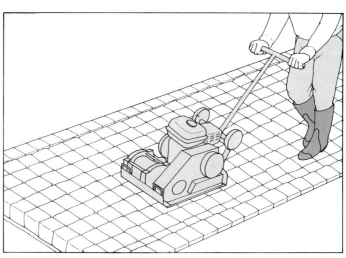

Move the vibrator slowly over the area two or three times, trying not to stop in any one spot for too long. Do not use

the vibrator within 1m of the edges where blocks are not yet laid: there is a risk of forcing the blocks sideways.

If the level after it has been compacted is still too high, try going over again with the vibrator. If it is too low you will need to lift the blocks and raise the level of the sand bed.

Once you are satisfied with the level, lay the remaining blocks or bricks and fill in with cut ones where necessary.

Then use the plate vibrator over the entire surface several times.

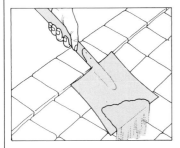

Spread a thin layer of fine sharp sand over the blocks and go over with the vibrator to force it into the cracks between the blocks. Keep sweeping more sand over and compacting until all the cracks are filled with hard, packed sand. Then sweep off the excess.

Laying bricks and blocks on mortar

Blocks or bricks laid on a 50mm bed of mortar are more stable than if laid on sand. As they will not be compacted as vigorously, the edging need not be quite as substantial, although if you are laying a drive a strong edge is recommended.

Prepare the sub-base as for laying on sand with marker pegs showing the thickness of the paving blocks or bricks

and a 50mm layer for mortar.

Access to the site is less important with this method, as you will be levelling each block as it is laid. However, make sure that you will not need to walk on the freshly laid blocks for 24 hours after laying.

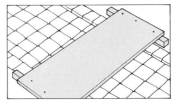

If it will be necessary to cross in the meantime, build a bridge with a wide board nailed to cross pieces of timber.

Mix a batch of sand and cement mortar – fairly dry, so that it just holds its shape when squeezed in the hand. Mix no more than enough to cover about two square metres at a time. The best arrangement is to have one person mixing while the another is laying and levelling the blocks.

To level the mortar to the correct height you can use either method described for levelling a sand bed (see pages 18-19).

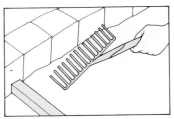

Shovel mortar onto the sub-base and use the back of a

rake to spread it around and settle it well into the gaps. Make sure that corners and edges are well filled. If you are using battens, scrape the excess off the top, remove the end one and trowel a little mortar into the gap. If you use a levelling board, tamp the mortar as if for a concrete slab, bringing the excess mortar towards the unfinished area.

Begin laying the blocks in a corner, tapping each one once or twice with the handle of the trowel to settle it on to the mortar. Leave a 10mm gap between blocks unless you are laying them in curves. In that case try to keep the widest joints not more than 25mm and the smallest not less than 6mm.

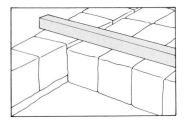

When you have laid a full row across the width, lay a batten across the marker pegs or resting on the edging to show which ones need tapping down to the right level.

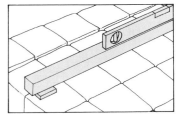

As you lay subsequent rows, use a batten lengthways to see

that the fall is constant and that the surface is flat in all directions.

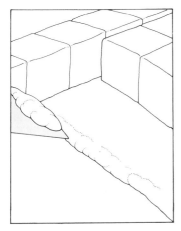

When you are within 200mm or so of the end of the mortar, lay the next bed, being careful to blend it well with the edge of the first.

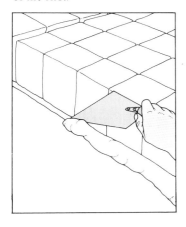

It is best to lay the complete area of mortar and blocks in one go but, if this is not possible, cut the edge of the mortar bed back to the last row of blocks leaving a neat vertical wall. Brush a mixture of 5 parts water to 1 part pva

adhesive on to the hardened wall just before joining fresh mortar.

When all the blocks are laid, leave them to cure for two days before filling the joints.

Pour a dry mixture of three parts sand to 1 part cement over the dry surface of the blocks and use a stiff broom to sweep it into the joints. Keep pushing more down the gaps and sweeping the excess off diagonally to prevent brushing the mixture out again.
When the joints are as full as possible, damp the entire surface with a fine spray from a hose.

Then quickly clean the surface by throwing sawdust or fine sand on to it and sweeping it off again with a soft broom.

If the mixture in the joints settles, repeat the operation when it has dried.

Laying paving slabs

Paving slabs are usually laid on sand or on spots of mortar. For a drive the slabs should be the hydraulically pressed type for strength. Cast paving slabs can be used for drives, but it is preferable to lay them on a 50mm bed of mortar.

The size and weight of slabs make edging unnecessary, although for a drive some reinforcment of edges is recommended.

To lay slabs on a bed of sand or mortar, follow the instructions given for bricks and blocks (see page 17).

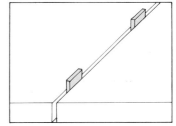

Lay sloping-edged slabs touching each other and square-edged ones with a 10mm gap all round. Use pieces of wood or plywood as spacers to keep the gaps constant.

To lay paving slabs on spots of mortar, prepare the sub-base as before and prepare a batch of stiff sand and cement mortar.

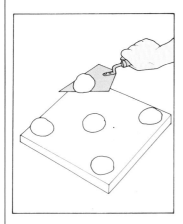

Before laying each slab, trowel cricket-ball-sized dollops of mortar on to the bottom face – one at each corner and one in the middle.

Place the slab in position right way up, and use the handle of a club hammer to tap it down gently to the correct level.

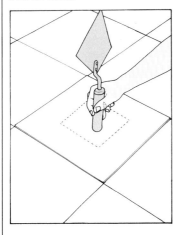

Try to adjust the slope as you tap it down by tapping it slightly off centre. Do not tap on the highest edge or corner or the slab will tend to rock. All the blobs of mortar should be compressed at the same

time to keep it stable.

It is important not to walk on the newly laid slabs for 24 hours, so begin laying in the least accessible area and build a bridge if traffic will need to cross during the drying period.

Laying crazy paving

Crazy paving slabs, whether of stone or concrete, are laid similarly to regular-shaped ones. It takes longer to fit them together in a satisfactory pattern and, because they are often of different thickness, more time is spent adjusting the sand or mortar beneath to keep them level.

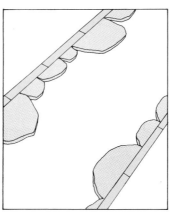

It is best to start by using the slabs with a straight side along the edges of the site. Space the larger slabs fairly equally apart and use medium and small ones to fill the gaps as you go.

Filling the joints is best done with sand and cement mortar and a trowel, especially for a drive or patio. For a path you can use a dry sand and cement mix swept into the joints. Because of the larger joints it will be necessary to repeat this process several times.

Laying sets

Sets are much thicker than slabs or blocks, so the sub-base will be lower. They are best laid packed close together on a mortar bed as described for bricks (see page 20). It is possible to lay them on a bed of sand but, as they have slightly irregular sides, the site needs to have very strong edges to hold them together. A plate vibrator is best for compacting them.

The joints are filled with a dry sand and cement mix as described earlier (see page 21).

Laying cobbles

Cobbles are usually laid on edge to discourage walking on particular areas, although they can also be laid flat to make a comfortable walking surface. Both ways are best on a mortar bed and both need edging to hold them together.

As patterns can be made by using both different sizes and different colours, keep them in separate piles or bags, and use buckets to hold the ones you are working with.

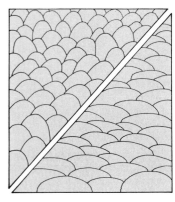

Press them into a bed of mortar, keeping the tops level.

If laying them flat, keep the flat sides up.

When the mortar has set, use dry sand and cement swept into the gaps. For cobbles on edge, fill the joints only about half full and spray with a fine mist of water. Cobbles laid flat need the joints filled, so several applications may be needed.

Laying pebbles and gravel

Pebbles (or pea shingle) or gravel can be laid loose over a normal sub-base. Like all loose paving materials, they require edges to keep them in place. These could be concrete or brick kerbs or a simple border of boards supported by stakes. Use pressure or vacuum treated timber.

The best finish is obtained by spreading a 25mm layer over the sub-base and compacting it with a garden roller. Throw a shovelful on to low areas and continue rolling until it is flat. The surface can be left like this or you can spread another 20mm or so layer onto the top and tidy with a soft rake.

Paving in a lawn

A practical and unobtrusive path can be made across a lawn by laying concrete slabs or stone flags in single or double rows. It is best to avoid a too regular arrangement. Odd-shaped slabs are often used as they can easily be formed into curves.

The stones or slabs should be laid flush with the surrounding soil so that lawn mowers can be used over them.

Begin by laying out the stones along the path to see how they will look and try walking on them to achieve the best spacing.

Use a narrow spade to cut through the turf around the edge of each stone. Then lay each stone to one side, keeping it facing the same way in which it will be laid.

Dig out each hole as for a sub-base. Allow about 100mm for hardcore plus the thickness of the slab.

Ram the hardcore down firmly and place dots of mortar as for the spot mortar method of laying paving slabs (see page 22).

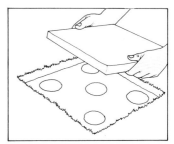

In this case it is easier to put the mortar on to the sub-base. Lower the slab into the hole and use the handle of a club hammer to compress the spots of mortar and bring the slab level with the soil. If necessary, lift the slab to add or remove mortar.

Repairs and Maintenance

Paving slabs

Paving slab repairs are simply a matter of removing the broken or sunken ones and re-laying. Use a spade to remove material from the joints and dig in under the slab.

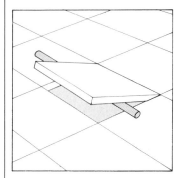

Lever it up enough to place a length of pipe or broomstick beneath it. Then roll the slab away from the hole.

If the slab was laid on a mortar bed, remove the old mortar and lay a fresh bed. If it was laid on sand, stir the sand and add a little more.

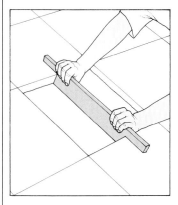

Level the new bed a few millimetres higher than the surrounding bed and lower the slab gently into the space.

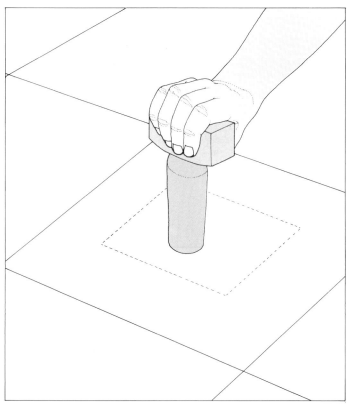

Use the handle of a club hammer to tap the slab level with the rest. This may require patience, as it may be necessary to lift the slab to add or remove some mortar or sand to get it right.

Use the same method with paving blocks or bricks.

Repairing concrete

Repairing concrete can be done with proprietary concrete filler or you can make your own. Mix a mortar of three to one of sand and cement and add a pva (polyvinyl acetate) adhesive to the water according to the manufacturer's instructions.

Repairing holes

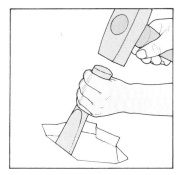

If you are repairing a hole, chop out the area to a depth of 12mm with a cold chisel and club hammer. This is the thinnest layer possible for the filler.

Brush the hole clear of dust and prime it with a solution of pva adhesive following the instructions on the can.

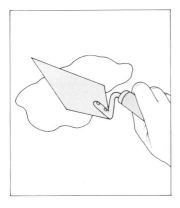

Press the filler firmly into the hole with a trowel and level off. Cover with polythene for three days.

Broken edges

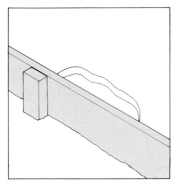

Broken edges of concrete slabs or steps can be repaired in the same way by using temporary formwork to hold the filler in place. A little petroleum jelly on the surface of the shutter board will ensure that the filler does not stick to it.

Remember to prime the edge

of the slab with adhesive solution. Cover the repair for three days before removing the formwork.

Resurfacing drives or paths

An old concrete drive or path can be resurfaced with asphalt or stone chippings held by a bitumen emulsion. The first step is to repair cracks and holes in the concrete.

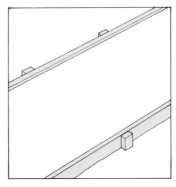

Set kerb edging or temporary wooden formwork along the edges 20mm above the surface of the concrete. Sweep the surface thoroughly.

Pour the emulsion over the surface and spread it with a

stiff broom. Follow the manufacturer's instructions as to the thickness to use. It should be thicker for stone chippings than for asphalt. Wait until it changes colour.

For asphalt, shovel or pour it over the surface and use a rake to spread it to a depth of 20mm. A batten across the edging will help with the final levelling.

Wet the drum of a garden roller to prevent the asphalt from sticking, and roll the entire surface. Add asphalt to depressions and roll again until all the asphalt is evenly compacted. A layer of stone chippings can be scattered over the asphalt before the final rolling, if desired.

To lay stone chippings directly on the emulsion, sprinkle just enough to cover the surface and roll immediately. The surface can be used right away, but wait a few days before sweeping off the loose chippings.

PATIOS

A patio should be easily accessible from the house so that the interior and the garden are brought together and living space is increased.

The best site for a patio is on the sunny side of the house, ideally with some protection from the prevailing winds. If the layout of the house is unsuited to this, try to plan the patio so that at least part of it catches the sun for the best part of the day. Shelter from the wind may be provided by shrubs in tubs or by a screen at the shaded end of the patio.

The materials and methods for making a patio are the same as those described for paths and drives. Because a patio is usually close to a house it requires especially careful planning (see pages 12-14).

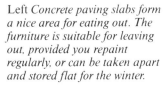

Left *Concrete paving slabs form a nice area for eating out. The furniture is suitable for leaving out, provided you repaint regularly, or can be taken apart and stored flat for the winter.*

Above *Crazy paving has been used on this city patio, providing a clean, flat area for tables and chairs. This patio furniture cannot be left out of doors in bad weather so, before buying it, make sure you have somewhere to store it.*

Above right *This barbecue area has been constructed at the back of a country house, forming a sheltered, sunny patio. Note how the green of the paintwork is echoed in the umbrella and cloth.*

Right *York stone flags have been laid between house and garden to provide a patio area that is sheltered, sunny and private. The mortar between the flags will keep the weeds out, but do not forget to allow for drainage.*

BUILDING STEPS

There are two ways of building steps in a path or patio. They can either be cut into an existing slope or built free-standing. Any paving material (except loose gravel) can be used for the treads. The risers are usually built of brick, but these can be covered with blocks or slabs and set in mortar. It is also possible to cast concrete steps using formwork. These can also have slabs, blocks or bricks over them.

First calculate the number and height of steps you will need. The general rule is that low risers should have wide treads and high risers, narrow ones. Garden steps should have risers of not less than 100mm and not more than 175mm.

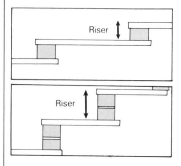

Use the examples above to calculate your steps by measuring the difference in height (rising) between the top and bottom of the steps.

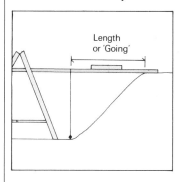

Use a spirit level and a timber batten long enough to span the

length of the steps. Place one end of the timber at the level of the top step and rest the other end on a step ladder or any support that is level. Then, hold a string with a weight on the end so that the weight is just touching the ground where the beginning of the bottom step will be. Mark the timber where the string crosses it. The distance from the mark to the end of the batten is the total length of the steps. This is called the 'going'.

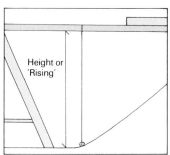

Measuring the height from the ground to the underside of the batten gives you the total height of the steps. This is called the 'rising'.

Use the examples above to calculate the number of steps and the length of each tread and the height of each riser by trial and error. Remember to include the thickness of the slabs, or whatever you will use to pave the treads, in the risers.

The easiest method of

building steps is to use paving slabs for the treads and bricks for the risers, but you can substitute any of the other materials following the instructions for laying them.

As far as design is concerned, the most important points are that the surface should be non-slip and the steps equally spaced.

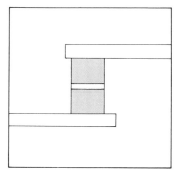

A good safety feature is to overlap each tread about 25mm over the edge of the riser below. This casts a shadow which makes the step easier to see in bad light. Also, treads should be sloped slightly, say 1 in 100, to ensure that rainwater does not stand on them.

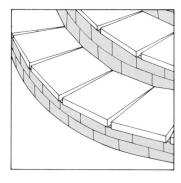

For curved steps, keep the curve as gentle as possible

(unless using curved blocks or bricks) to avoid large gaps between slabs. If the curve is tight use smaller sizes of slab to keep the front edge of the treads from appearing jagged and to keep the joints smaller.

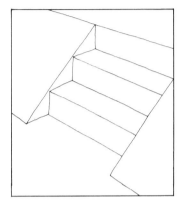

Steps in a slope are the easiest to build as the earth can be cut roughly to shape and each tread can be laid on a separate sub-base with each riser built on the back of each tread.

The first riser will need a concrete sub-base unless it is supported by a path or some other paving.

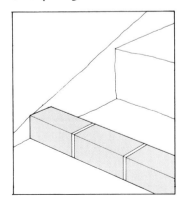

Cut the steps in the slope leaving 100mm for the

hardcore plus 50mm for the mortar bed plus the thickness of the paving.

Build the first riser laying the bricks on mortar and leave for at least two hours to set.

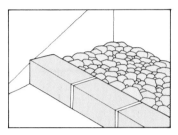

Fill the step behind the riser with hardcore and ram it down, being careful not to disturb the freshly laid bricks. Blind the hardcore with ballast and lay the slabs on a 50mm thick bed of mortar. Remember to slope the step for drainage.

Build the next riser the correct distance back from the front edge of the slabs and repeat for each step.

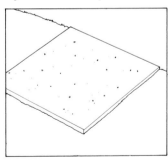

Free-standing steps need to be supported by brick walls built up from the lowest level.

The simplest method is to lay a 75mm thick concrete slab on a sub-base covering the entire area of the steps. Cover it and allow three days for it to set.

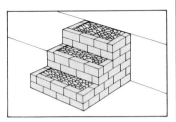

The risers and walls are then built to the correct heights and the centre filled with hardcore 40mm below the tops of the risers. Then the treads are laid on a 50mm bed of mortar.

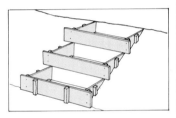

Concrete steps are made from the bottom up, using formwork to lay each step on the one beneath. Use a normal sub-base and build the formwork to the required heights. Nail the boards across the front of the risers and brace the sides of the formwork well.

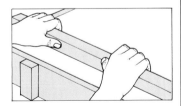

Pour the first slab and level off the tread portion. Delay pouring the next slab until the concrete is firm enough not to deform under the weight.

Leave the finished steps covered for three days before removing the formwork.

Unusual Ideas

Left *This patio has been constructed from old bits of rock and stone left lying around the house. Plants have been encouraged to take root between the stones and the addition of wooden furniture completes the rustic effect.*

Bottom left *Attractive patterns can be achieved with stone of different sizes. Here the design incorporates many small bits of stone laid between the larger pieces.*

Below *Different coloured concrete paving slabs, with spaces left for plants, make an attractive, low-maintenance front garden.*

Top right *Riven stone has been painstakingly laid to form a dry-stone path. It has the attractive effect of flowing rather like lava.*

Middle right *Space for plants was left when this cobbled area was laid, and also helps with drainage.*

Bottom right *Wooden blocks of varying sizes have been let into the lawn to make an unusual but visually pleasing path. Care must be taken as wooden paths can become slippery if they are wet for extended periods. They are therefore not suitable for areas of shade, but should be out in the open.*

TOP TEN TIPS

1. Remember when choosing paving slabs that smooth slabs can be slippery when wet.

2. When mixing concrete or mortar, always keep one shovel and bucket dry to be used for the cement. Any moisture will contaminate the bag of cement.

3. To prevent weeds from growing through gravel and loose paving, treat paths and drives in spring and autumn with a long-lasting weedkiller.

4. Moss makes paving slippery. Use a proprietary fungicide or a solution of household bleach on patios and steps to prevent it.

5. Set paving in a lawn flush with the surrounding earth so that lawn mowers can be used over them.

6. If you build a sandpit for children in a patio, make a cover for it to keep out rain and local cats.

7. A sandpit makes a good site for a pond after the children have grown up.

8. To avoid too much wear on grass at the foot of steps, lay an extra tread or some stone flags into the lawn.

9. Plant Spearmint and other sweet-smelling plants between the paving slabs on your path. Plant to one side, avoiding the centre area of heavy traffic, where the plants will not survive.

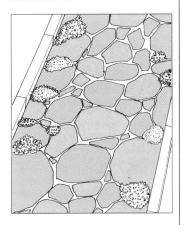

10. Lay two colours of slab in a chequerboard pattern to form a giant chess or draughts board; use painted flower pots as chessmen.

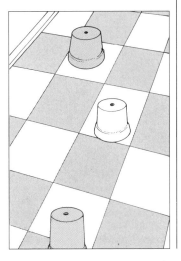

Author
Dek Messecar
Series Consultant Editor
Bob Tattersall
Design
Bob Lamb
Research
Liz Whiting & Amalia Grassi
Editor
John Stace
Illustrations
Rob Shone

Dek Messecar is a professional joiner who has had experience on all
aspects of DIY.

Bob Tattersall has been a DIY journalist for over 25 years and was
editor of *Homemaker* for 16 years. He now works as a freelance
journalist and broadcaster. Regular contact with the main DIY
manufacturers keeps him up to date on all new products and
developments. He has written many books on various aspects of DIY
and, while he is considered 'an expert', he prefers to think of himself as
a do-it-yourselfer who happens to be a journalist.

Photographs from Elizabeth Whiting   Cover Photography by Carl Warner
Photo Library   Materials provided by
                                       Erith Building Supplies

The Do It! Series was conceived, edited and designed by Elizabeth
Whiting & Associates and Robert Lamb & Co
for William Collins Sons & Co Ltd
© 1989 Elizabeth Whiting & Associates and Robert Lamb & Co

First Published 1989
9 8 7 6 5 4 3 2 1

ISBN 0 00 412 475 8

Published by William Collins Sons & Co Ltd
London · Glasgow · Sydney · Auckland
Toronto · Johannesburg

Colour separations by
Bright Arts, Hong Kong.
Printed in Spain